YOUR KNOWLEDGE HAS VALUE

- We will publish your bachelor's and master's thesis, essays and papers

- Your own eBook and book - sold worldwide in all relevant shops

- Earn money with each sale

Upload your text at www.GRIN.com and publish for free

Manipulating Microscopic Objects. A Study on Optical Tweezers and Acoustic Levitation

Jefferson Martin

Bibliographic information published by the German National Library:

The German National Library lists this publication in the National Bibliography; detailed bibliographic data are available on the Internet at http://dnb.dnb.de.

ISBN: 9783346846983
This book is also available as an ebook.

© GRIN Publishing GmbH
Nymphenburger Straße 86
80636 München

Print and binding: Books on Demand GmbH, Norderstedt, Germany
Printed on acid-free paper from responsible sources.

The present work has been carefully prepared. Nevertheless, authors and publishers do not incur liability for the correctness of information, notes, links and advice as well as any printing errors.

GRIN web shop: https://www.grin.com/document/1340498

Manipulating Microscopic Objects: A Study on Optical Tweezers and Acoustic Levitation

Marius Martin

Content

Abstract

This thesis investigates two powerful techniques for manipulating and studying microscopic particles: optical tweezers and acoustic levitation. Optical tweezers use the force of light to trap and manipulate small particles, while acoustic levitation uses sound waves to levitate particles in mid-air. The first part of the thesis focuses on the construction and optimization of an optical tweezers system. The system is build using high-powered laser, optics for focusing the laser beam, and a position detection and control system. the second part of this thesis explores the use of acoustic levitation for the manipulation of micro spheres. The acoustic levitation system is designed and constructed using 72 transducers. The system is characterized using measurements of the acoustic field and the resulting levitation forces. The levitation of micro spheres is demonstrated. This thesis would provide a detailed investigation of two powerful techniques for the manipulation of microscopic particles.

Introduction Optical tweezers

Optical tweezers and acoustic levitation are two powerful techniques used to manipulate and study microscopic objects in physics, chemistry, biology, and engineering. While they have different operating principles and applications, both techniques are based on the control of forces exerted by light and sound waves on small particles. Optical tweezers use a highly focused laser beam to trap and move tiny objects, such as cells, bacteria or beads, in three dimensions. The laser beam creates an optical gradient force that attracts the object towards the centre of the beam, and a scattering force that pushes the object away from the beam axis. By manipulating the position, intensity and polarization of the laser beam, researchers can precisely control the position, orientation, and the shape of the trapped object. Optical tweezers have revolutionized the field of biophysics by enabling the measurement and manipulation of molecular forces, the study of cellular mechanics and the development of novel nano-materials. On the other hand, acosutic levitation uses ultrasonic sound waves to levitate and suspend small objects in mid-air. The sound waves create a standing wave pattern that produces acoustic radiation forces that can balance the gravitational force on the object. By adjusting the frequency, amplitude, and phase of the sound wave,researchers can control the position and movement of the levitated object. Acoustic levitation has been used to study the properties of materials in microgravity, to synthesize and manipulate micro- and nano particles, and to develop novel drug delivery systems. Both optical tweezers and acoustic levitation have advanced our understanding of the behaviour and properties of microscopic objects and have opened up new avenues for research and applications. By combining these techniques with other tools. Such as microscopy, spectroscopy, and computer simulations, researchers can gain unprecedented insight into the complex and fascinating world of the nanoscale.

Theory

The principle behind optical trapping is based on Hooke's law, which states that the optical force moves the trapped particle in a certain displacement x to the equilibrium position, and the trapped particle would experience string stress during the process, in this case, we call it stiffness k. The trapped particle on the sample plane would be controlled by a scattering force in the upward direction and downward gravitational force, the gradient force would stand orthogonal to the direction of propagation. The gradient force is proportional to the gradient of intensity and towards the direction of the intensity gradient, whereas the scattering force is proportional to the direction of the optical intensity and towards the direction of incident light. The equations of those two forces can be given as

$$F_{scatt} = \frac{I_0 \sigma n_m}{c}, \quad \sigma = \frac{128\pi^5 a^6}{3\lambda^4}\left(\frac{m^2-1}{m^2+2}\right)^6$$

$$F_{grad} = \frac{2\pi\alpha}{cn_m^2}\Delta I_0, \quad \alpha = n_m^2 a^3\left(\frac{m^2-1}{m^2+2}\right)$$

eq[1].

Because of the purpose of calculating the optical force by using the equation of Hooke's law, the factor of stiffness would be required to be determined. The most common methods are power spectrum density(PSD), auto-correlation analysis(ACA), and potential analysis(PA) which require time series during the measurements. However, the limitations are that the execution time would take longer and obtain more relative errors. Another method is called maximum likelihood analysis(MLA), it has the advantages of less sample needed, easier to be manipulated based on the algebraic algorithm, less execution time and a relative error can be reduced to 2%. Additionally, MLA can also measure the conservative and nonconservative force field in 2 and 3 dimensions, it can track the stable, unstable, and saddle points around the equilibrium position.

Construction of optical tweezers

The mechanism of the optical tweezers is to use the laser beam to trap the tiny particle without contact. First of all, the laser beam would come through the objective telescope to enlarge the beam size (the telescope would be arranged in the 1:1 ratio, so the laser beam comes to the first lens to form a focal point, and then be enlarged after the second lens). The enlarged laser beam is redirected by the dichroic mirror to the objective lens, which uses high magnification 100× oil immersion with a numerical aperture of 1.32 NA(the high magnification can work well to form a sharp focus on the particle). The trapped particle would be presented in the middle of the objective and condenser. The condenser is used to collect the scattered light. The beam would reach the quadrant photodiode after being redirected by the second dichroic mirror. The quadrant photodiode can measure the position and the optical forces. Illumination light would bring the image to the CCD camera.

Choosing Laser source is the first step to constructing optical tweezers system, the common choice is a high-powered, continuous-wave laser with a wavelength in the near-infrared region,such as Nd:YAG laser.Optical tweezers would require laser beams with wavelengths at 532 nm and 1064nm. Laser beam should be powerful enough to create a strong gradient force and to trap small particles in order for trapping the aerosol particles, the studies on investigating the silica particle would make use of the laser beam(diode-pumped solid-state laser DPSS) with a wavelength of 532 nm.[1] One of the reasons for using this type of laser is its lower cost. In addition, it could lower heating and the wavelength located at the minimum of the absorption curve of the liquid.[2] Laura P. et al.(2018) indicated that they applied the laser beam with 532 nm to trap the silica microsphere in the liquid solution while the power is at 0.8 mW.[3] 532nm laser beam was applied with a quadrant photo-diode detector to discover more qualitative measurements. [4] Then the laser beam is directed through a series of optics, including lenses and lenses and mirrors to focus it into a small spot. The focal point of the laser beam is where the optical trap is created. The size of the spot is determined by the properties of the lenses and the laser wavelength. Sample chamber is where the particle are introduced and trapped. The chamber should be carefully designed to minimize turbulence and to prevent particles from sticking to the walls. Position detection system is used to detect the position of the trapped object. It involves using a quadrant photo detector or a camera to monitor the scattered light from the trapped object. The position of the object can be determined by analysing the position of the scattered light. Control system is used to adjust the position of the laser beam in order to trap and move the object. It can be done using a feedback loop that adjusts the position of the laser based on the position detected by the position detection system.

Method

The experiment would be provided with some analysis approaches to determine the stiffness of trapping. Potential analysis, auto-correlation (ACF) method, and power spectrum density method (PSD) have been applied to determine the trap stiffness. For instance, potential analysis is a method that is directly applied the equipartition theorem to measure the positions of distribution of particles.[5][6] In this method, the Boltzmann distribution would be applied to the equipartition method, the stiffness factor could produce Gaussian distribution if the potential is harmonics. The power spectrum density method (PSD) is based on the harmonic trap with the use of particle trajectory[6]. The relation of finding PSD requires determining the diffusion coefficient constant and friction coefficient, then the stiffness factor could be obtained. In general, a time series of correlated particle positions in a time interval would be needed to measure the PSD, which is the same for the ACF method. For the method of ACF, it needs to consider the Boltzmann constant and absolute temperature in order to measure the stiffness of trapping. A.Magazzu et al. (2022) showed that they used ACF and PSD methods to determine stiffness for single trapping grain by calculating the relaxation frequencies of the trapped particles. In this study, they stated that the stiffness k in three dimensions increases with the laser power linearly.[7] They asserted that the stiffness is asymmetric in the x and y directions, and propagates in the direction of z. The Gaussian beam would lower the stiffness in the z-direction along the z-axis. Additionally, they obtained the stiffness efficiency qi in three dimensions corresponding to the laser power. They obtained the stiffness efficiency in the x direction is 1.38, in the y direction is 1.104, and in the z direction is 0.536 with the unit of pNum^1mW^1. D. McGloin et al. (2008) showed that measuring the stiffness both in water and in the air for the silica particle was done by using the power spectrum method with detector sensitivity β for both over-dumped and under-dumped cases. They trapped the silica particle diameters at 1.86 and 3.01 um. The particle size of 1.86um has a trap stiffness of 0.88 ± 0.04pNum^-1mW^-1 in water, and 2.02 ± 0.08 pNum^-1mW^-1 in air. The particle size of 3.01um has trap stiffness of 0.49 ± 0.02pNum^-1mW^-1 in water and 1.50 ± 0.05pNum^-1mW^-1 in air. [4] The results showed that the trapping silica particle can be accomplished in both liquid and air with such particle sizes and values of power. Those methods are based on the measurements of obtaining the Brownian fluctuations of the particle surrounding the stable equilibrium position.

Those approaches have been applied for the measurements of trapping stiffness and determination of optical forces, which have good agreement on the efficient execution and reliable results. Researchers still have worked on finding better approaches to determine the stiffness of the trapping and make more efficient manipulation with less error. The approach called the Maximum likelihood estimator was investigated by Laura P. et al.(2018) [6] who stated that the stiffness k in 1 dimension by applying the maximum likelihood estimator (MLE) corresponds to the dependent and independent variables. They showed the calculation of the overdamped Langevin equation and average viscous friction force, which could be used by MLE to determine the stiffness of trapping and the diffusion constant. The results were performed by figures and can be found in[6], and compared to the Auto-correlation(ACF) and power spectrum (PSD) methods. It showed that the MLE method could get a relative error of 2%, which is much less than the 20% relative error obtained by the ACF and PSD methods. Laura P. et al.(2018) declared that the benefits of MLE are more than the methods of ACF and PSD. The MLE requires fewer samples and is easier to be executed on Matlab due to the linearity-based algorithm, the relative error is much smaller. Since the MLE method required the time series, the limitation of executing the MLE would be caused by the magnified blur images in the velocity spectrum. [8] One way to fix this situation is to convert the position of the trapped particle due to the exponential trajectory of the trapped particle at, H; S.A.B.A.D.J.R.-D. (2021) indicated. Laura P. et al.(2018) showed the approach of MLE in 2D can obtain the non-conservative force field by applying the spin angular momentum (SAM) and argued that the sophistication of determining the exact values of SAM even in the focused light beam due to the properties of the particle.[6] They made use of the Langevin equation in the case of over-damped harmonics and average friction force in the time interval, then the rotational matrix with symmetric and antisymmetric for the non-conservative force field. As Laura P. et al. mentioned, due to its tiny size, the non-conservative force field can not be measured by calculating the rotations in the direction of the beam axis. Finally, they firmly proved that the negative circular polarisation allows the OAM and SAM to have the same rotation of the particle. Laura P. et al.(2018) stated that the MLE in 3D can be used to detect stable and unstable points at the equilibrium position. They indicated that the negative unstable equilibrium points in the harmonics potential take advantage of determining the properties of the force field surrounding the equilibrium points. The method of MLE applied in 2D and 3D is required to make use of the spatial light modulator (SLM) in the plane. The 2D measurement can determine the conservative and non-conservative force field by comparing the OAM and SAM in the focused laser beam. In 3D, the approach of MLE can be used to study the stable and unstable points around the equilibrium positions.

In the scattered field, the trapping can be treated either as the particle passing the greatest field due to reduced dipole moment, or a change in momentum by the divergence of the transmitted beam at the low refractive index. Therefore, the reaction force would be obtained either by obtaining the values of the electric multipole moment or the change in momentum. In this case, the approach of the transition Matrix which was introduced by waterman would need to be applied in measuring radiation forces and torques. Transition Matrix (T-Matrix) makes use of the formulation of the integration of electromagnetic waves over a spherical multipole field based on the Maxwell stress tensor. A.Magazzu et al.(2022) asserted the method of using a T-Matrix in the investigation of non-spherical shaped dust grains to measure the optical forces,[7] and obtained coherent results for both theoretical and experimental. This evidence claimed that the T-matrix can be used to measure irregularly shaped particles by calculating the torque of the rotations.

T-Matrix is independent of the direction of propagation and the polarisation states and the scattered field; it can precisely determine the optical forces and torques. T-Matrix is executed on the Matlab software in order to investigate scattering properties. T.A. Nieminen et al.[13]stated that the computation can be manipulated repeatedly for multiple incident field cases after the T-Matrix algorithm is calculated, which would need to apply matrix-vector multiplication by determining a new vector with another incident field. In addition, the arbitrariness of determining the scattering field and incident field is another principle of calculating the T-Matrix, which means that it would work by using any approach to determine the scattered field and forces. Therefore, those two features of T-Matrix are good for modelling optical tweezers. P. W. Barber & H. Massoudi (2007) [14] asserted that the T-Matrix method could be used for studying the resonance region scattering and absorption characteristics of arbitrary-shaped dielectric particles. In addition, the linearity-based formulation of the T-Matrix corresponding to the scattered field of unknown coefficient and incident field of known coefficient could be obtained from the relation of spherical harmonic expansions of the incident and scattered fields to the boundary condition crossing the surface of the spherical particle.

The formulation of the T-Matrix is called the extended boundary condition method (EBCM) which is widely used and can be treated as Mie theory. The benefit of applying the EBCM is that it can be used in 2D methods by integrating only surface integral. In the 1D method, it can be used for axially symmetric particles. [13] However, the limitation of EBCM is that it limits the measurement of isotropic and homogeneous particles. Furthermore, the particle would experience exceptional changes in shape when executing the computation. In addition, T.A. Nieminen et al.(2010) [13] indicated that the EBCM limits the computation of the star-shaped particle. Lorenz-Mie solution is another method for the T-Matrix, the advantage of it is that its computation is easier. EBCM. P. W. Barber & H. Massoudi (1982) [7] claimed that the forward Lorenz-Mie solution can be used to determine the size and the refractive index for both spherical and non-spherical particles. The problems of trapping in the light scattering would be caused are that the trapped particle would be either excessively large for the accuracy of Rayleigh approximation or small for ray optics. Additionally, T.A. Nieminen et al. stated[13] that the incident illumination is not a plane wave, but a focused Gaussian beam.

Mie's theory states that the boundary condition is the balance between external and internal electromagnetic fields on the surface of the sphere. Based on the Mie theory, the nanosphere particle can be determined by computing the summation of the harmonics' potential functions. However, the limitation of the Mie theory is that it is an analytical-based tool dealing with the plane wave illuminated by spherical particles. The incident light scattering would produce a focused Gaussian beam which would result in degenerating the accuracy of the computation.[15] The discrete dipole approximation (DDA) method is another way of determining the optical forces of dielectric particles in the electromagnetic field.[16] The advantage of the DDA method is that it can compute accurately for both regular and irregularly shaped microparticles in the trap. When analysing arbitrarily shaped particles, the method of DDA would be useful for calculating T-matrix. The benefits of the DDA method are that it only takes into account the volume of the discretized particle while leaving the surrounding medium. [17] Its execution is good at low-contrast scatterers and easier to formulate T-matrix. The discrete rotational symmetry of the trapped particle can be applied through the method of DDA to decrease the execution conditions during computation. Furthermore, the DDA method with point matching can be used to efficiently compute reliable results. Timo A. Nieminen et al. (2008) [17] demonstrated that the method of both the far-field and near-field point matching was exploited by applying the VSWFs field to obtain the DDA fields.

Review

The size of the trapped particle depends on the radiated power of the laser beam, the high-index particle, and the polarizability, which is for the purpose of the particle remaining in the state of being trapped. Challenges would be concerned when the trapped particle in the Rayleigh and Mie regimes. The observation of silica particles in water by using a single beam with gradient force has been investigated by A.Ashkin with their group. A.Ashkin et al. (1985) [9] showed that the trapped particle would be diminished in size and vanished after being trapped for ten minutes for the diameter of a 1.0um sphere with the power of a milliwatt.The size of a 0.173um sphere could be trapped in a few minutes by applying the same power. However, the diameter of the particles at 0.109um would perish very rapidly by using the power in the range from 12-15mW. It would even be more sophisticated to hold the diameter of particles at 85 or 38 nm. Additionally, they indicated that the minimum particle size of 26 nm has a scattering factor of 3e4 with an applied power of 1.4W, which is less than that for the 0.109um. They concluded that diameters of the particle size of 19.5 nm and 12.5nm with the high-index particles of 1.20 and 2.26 can be observed at the same power. Based on the evidence provided by A.Ashkin with their work, the trapping of a single beam by applying gradient force could be accomplished in the full-size range of Rayleigh to Mie regimes. Another study on trapping size particles in the air required to determine the axial trapping efficiency Q. D.McGloin et al. (2008) [4] showed that trapping silica particles in the air can be accomplished in the sample chamber, the particle having sizes of 1.86um and 3.01um could be trapped at continuous power in several minutes before vanishing. They showed the diameter 1.86um has minimum trapping power at 0.1 ± 0.02 mW and axial efficiency 0.19 ± 0.04. The diameter 3.01um has minimum trapping power at $0.29 \pm$ 0.03mW and axial efficiency0.28 ± 0.03. In addition, they also showed that a silica sphere with particle sizes of 4.32um produced by nebulizer compAir could be trapped simultaneously at power 3mW and disappeared. They found the limitation that the trapping could be unstable in this case. They indicated particles trapped in water with particle size 1.86um has a trapping power of 25mW, the particle size 3.01um has a trapping power of 130mW. This study showed that the trapping powers in the air are less than that in water. Comparing the work A.Ashkin has done on observing the size of particles, and D.McGloin with their group studied on observing the larger particle size both in water and in air. The above studies showed that it could be difficult to trap the smaller particles. The measurements of the axial trapping efficiency have been obtained by Laura M. and Jonathan P. R. (2009), which showed values of trapping efficiencies 0.1 ± 0.04 in liquid and 0.45 ± 0.12 in air with applied power around 1mW.[10] Those results measured under applying lower power are acceptable compared to the work done in[4]. Additionally, the displacement of the trapped aerosol particle surrounding the equilibrium position has been detected by Di Leonardo et al.(2007) and showed the transformation was presented between the low-powered under-damped harmonic regime and the high-powered over-damped harmonics regime.[11]

Stability

In order to form a stable trap, the light beam needs to be focused, then make a three-dimensional intensity gradient.[5] Bouloumis, T.D.(2020) demonstrated that making use of Fresnel coefficients with a counter-propagation beam results in the cancellation of the scattering forces to form a stable trap.[12] Specifically speaking, the incident angle with a degree larger than 70 would focus the particle firmly under the condition of a high numerical aperture of the objective lens. The stable trap would be accomplished when the gradient forces overcome the scattering forces after the surface of the particle is exerted by the ray with a small incident angle. On the other hand, if the incident angle is less than 30 degrees, single beam trapping would be failed and the cancellation of scattering forces with the same characteristics would be led by using two counter-propagating beams, which is another way of creating a stable trap. Ashkin et al. (1986) showed the approach of forming a stable trap with energy 10KT in the harmonics potential well holds out against the Brownian motion in liquid.[9] As shown by the equation of the gradient force, the intrinsic polarisability is proportional to the radius of the particle to the power of three times. Ashkin demonstrated their work on creating a stable trap, which is achievable when the trapped particle has a larger size, in order for the gradient force exerted on the particle. On the contrary, in the situation of trapping smaller size particles, the gradient force would be reduced three times more than the radius of the trapped particle, which shows that the trap of the particle exerted by gradient force would be unstable.[12] Based on the evidence above, the size of trapped particles is key to creating a stable trap, the incident angle is also needed to take into account for producing a stable trap.

Introduction to acoustic levitation

This part of the thesis investigates the current scientific knowledge on acoustic levitation with a focus on potential applications. Acoustic levitation is a method of suspending an object in the air through an acoustic field, and offers an intriguing prospect for future use, with potential advancements in manufacturing, transportation, pharmaceuticals, and other areas.Acoustic levitation is a technology that uses sound waves to suspend and manipulate objects in mid-air. It has a variety of modern applications in industry, medicine, and research, and its use could potentially revolutionize many current technologies. In this thesis, the concept of acoustic levitation, its practical applications, and its limitations will be explored. Levitating the small particles using standing waves is one approach, which consists of acoustic radiation force. The technique under this theory to trap small objects is called The acoustic levitator, which levitates solid and liquid using acoustic radiation forces. It is categorized into several classes, which are the single-axis standing wave, multi-axis, near field, and single beam levitation. Specifically, Single-axis is the simplest and the most common to be used to levitate objects.[18] Acoustic levitation uses sound waves to trap the particle vertically in the mid-air, the trapped particle would be levitated in the middle by the transducer and reflector creating a pressure node[20]. The sound wave would bring the momentum to the trapped particle to form a stable trap by the acoustic radiation force.[19] Distinguish optical trapping and acoustic trapping is that multiple material objects with different sizes can be levitated by the acoustic levitator, while optical trapping accomplishes the trapping of particles on a micro scale[19]. As mentioned above, single-axis levitation is the simplest approach, while it involves two different types of designs. The first one contains a transducer and reflector to generate resonance in the cavity, alternatively, the non-resonance levitator could be constructed by two emitters facing each other[19]. Two of those levitators generate sinusoidal excitation signals with the standing waves; the trapped particles would be levitated under the nodes created by the standing waves. The acoustic force can be large enough to levitate objects against gravity, the acoustic levitation uses sound waves to levitate material with ultrasonic frequency in the range of greater than 20kHz. [28]

Standing waves in acoustic levitation
Standing waves are the key to creating the acoustic forces that enable acoustic levitation. A standing wave is a type of wave pattern that is created when two waves of the same frequency and amplitude are travelling in opposite directions and interfere with each other in a way that causes the wave to appear to be stationary. In an acoustic levitation system, standing waves are created by reflecting sound waves off a hard surface such as a reflector or a resonator. The reflected waves combine with the incoming waves to create high and low-pressure areas. When the object being levitated is placed at a certain point within the standing wave, it experiences an upward force that balances the downward force of gravity and allows it to remain suspended in mid-air. There are two types of standing waves that are commonly used in acoustic levitation: longitudinal standing waves, and transverse standing waves. Longitudinal standing waves are created when sound waves are reflected back and forth between two reflectors or resonators that are placed parallel to each other. The resulting standing wave has areas of high and low pressure that run parallel to the direction of wave propagation. Longitudinal standing waves are commonly used in single-axis acoustic

levitation systems. Transverse standing waves are created when sound waves are reflected back and forth between two reflectors or resonators that are placed perpendicular to each other. The resulting standing wave has areas of high and low pressure that are perpendicular to the direction of wave propagation. Transverse standing waves are commonly used in multi-axis acoustic levitation systems. In both types of standing waves, the frequency of the wave is determined by the distance between the reflectors or resonators and the speed of sound in the medium the amplitude of the wave can be controlled by adjusting the intensity of the sound waves produced by the transducers. Therefore, it is essential to understand the standing waves used in acoustic levitation must be carefully controlled and stabilized to maintain the levitated object in a stable position. Any disturbances to the standing wave can cause the object to become unstable and drop out of the levitation area. Therefore precise control of the transducer and reflectors is important to achieve stable acoustic levitation.

In a single-axis acoustic levitation system, the levitation object is trapped in a standing wave created by a single ultrasonic transducer. The standing wave is characterized by regions of high and low pressure, which exert forces on the object in both the axial and radial directions. Therefore, there are two forces are concerned, axial force and radial force. The axial force in a single-axis acoustic levitation system is the force exerted by the standing wave in the direction of the sound wave propagation. This force is proportional to the gradient of the sound pressure along the axial direction, and it is typically strongest at the pressure nodes of the standing wave. At these nodes, the pressure variations are maximal, and the object experiences a maximum axial force that is directed towards the center of the node. The radial force in a single-axis acoustic levitation system is the force exerted by the standing wave perpendicular to the direction of the sound wave propagation. This force is proportional to the gradient of the sound pressure along the radial direction, and it is typically strongest at the pressure antinodes of the standing wave. At these antinodes, the pressure variations are maximal, and the object experiences a maximum radial force that is directed towards the ccenter of the antinode. In a single-axis acoustic levitation system, the axial and radial forces ar etypically not equal, and they depend on the size, shape, and material properties of the levitated object, as well as the frequency and intensity of the ultrasonic transducer. The axial force is generally stronger than the radial force, which can cause the levitated object to oscillate orr vibrate in the radial direction. To achieve stable levitataon, the axial and radial forces must be carefully balanced.

Configuration of acoustic levitation
The theoretical work would require an actual attempt to prove the results. Acoustic levitation consists of two types of levitating, which are standing wave levitation and squeeze film levitation. Standing wave levitation is the simplest levitation which is also called far-field levitation, this can be done by the method called free-field piston source. Squeeze film levitation is near field levitation. In this case, I will describe the standing wave levitation. In order to construct the acoustic levitator, the emitter consists of transducers, so that it can help produce a standing wave with angular frequency at the bottom. The top reflector would stand in the opposite direction to the emitter. Both the emitter and reflector consist of 36 transducers and are arranged in the shape of hexagons 6, 12,18 orders. There would be a camera replaced in the center of the transducers to record the image.

Set up of acoustic levitator

Transducers

The transducers are a crucial component in acoustic levitation systems. They are responsible for generating the sound waves that create the standing waves used to levitate objects. The most common type of transducer used in acoustic levitation is the piezoelectric transducer. Piezoelectric transducers are made of a material, such as quartz or lead zirconate titanite (PZT), that exhibits the piezoelectric effect that occurs when a voltage is applied to the material causing it to vibrate and produce sound waves. The frequency of the sound waves is determined by the thickness of the transducer and the speed of sound in the material. Piezoelectric transducers can be designed in various shapes and sizes, including discs, cylinders, and rectangular plates. The shape and size of the transducer depend on the specific requirements of the levitation system. For instance, a disc-shaped transducer may be used in a single-axis levitation system, while a rectangular plate may be used in a multi-axis levitation system. Additionally, piezoelectric transducers and other types of transducers can also be used in acoustic levitation, such as magnetostrictive transducers using the magnetostrictive effect to generate sound waves. This effect occurs when a magnetic field is applied to a magnetostrictive material, causing it to change shape and produce sound waves. However, piezoelectric transducers are generally preferred due to their higher efficiency and ease of use.

Components
1. Transducers (76)
2. 3D printer base
3. Arduino nano
4. DC-barrel
5. L298N Dual Motor Drive Board
6. Black and red wires
7. DC adaptor (9V)
8. Switch
9. Jumper wire

Other things needed for building acoustic levitator
1. Multimeter
2. Oscilloscope
3. Computer
4. Soldering iron
5. Glue

In order to construct an acoustic levitator, first of all, is to check the polarity of the transducers by using either an oscilloscope or a multimeter. Although polarity is labeled already on each of the transducers, they may be not in the correct order. Polarity is in the DC circuit, which makes current flow from the positive terminal to the negative terminal. There are a bunch of ways to test the electric polarity. Oscilloscope is one way to test the polarity by identifying the directions of the spikes, Arduino can also test the electric polarity by signals, and using the multimeter is another way to test the electric polarity, which is the simplest way for testing polarity.

The testing polarity of the transducers using a multimeter is the approach that would be used in the experiment. There are 76 transducers, each of them having around 43kHz frequency emitting sound waves, and one positive and one negative terminal(but only 72 transducers would be used in the construction, so there are 4 spare ones). Connecting the red probe wire to the input on the multimeter and the black probe wire to the ground. When the reading of the transducer shows negative values, then mark another terminal as positive. Each of the transducers would show around 17 volts on the multimeter, after testing all 76 transducers, there are 75 works and one is malfunctioning. The next step is to glue all of the transducers to the 3D-printed base. They are shaped as hexagons and in the order of 6, 12, and 18 from inside to outside. In order to wire the transducers in the right direction, the positive terminals are put in the inner circle, and the negative terminals are put in the outer circle. This arrangement would make wiring up easy. The next step is to glue them tightly to the 3D-printed base. Red wires are provided to connect all of the positive terminals, black wires are for those negative terminals. The Soldering Iron can help to connect them stably. Now two long wires are needed to connect the positive and negative terminals separately to the drive board. There are black and red wires on the top array and two black and red on the bottom array. Then connect red wires and black wires to the driver board, (make sure they are in the correct order). Then the process reaches the step of programming the Arduino, the code can be found on this website[31]. After downloading the Arduino IDE and the code on the laptop, then click the tools to set the board as "Arduino Nano", select the correct COM port, circle both of the compile and skitch and compile in the preference (these steps are important, otherwise the code would not be uploaded successfully). Connect the USB cable to the Laptop and upload the installed code to Arduino Nano. Jumper wires are used to link between the Arduino and driver board, which four female jumpers connect from Arduino (A0,A1,A2,A3) to the inputs of the driver board (IN1,IN2,IN3,IN4). There is another male-female jumper wire connecting to the Arduino ground(D2,D3 or D4), which is used to control the position of the levitated particle. The DC supply needs to be connected by using black and red wires, then the black wire on the DC supply connects to the switch, a red wire connects to the 12V on the driver board. The ground on the switch is connected to the middle connector of the driver board and the ground on the Arduino Nano(inserted a male-female jumper wire). Another male-female jumper wire is used to connect 5V on the driver board and 5V on the Arduino Nano, then the screwdriver tights them up.

After making the acoustic levitator work, several tests can be done by levitating lighter objects, such as polystyrene. This procedure is done for the purpose of the levitation function. If the levitation could be done successfully, then the experiments can be run with multiple material levitation, and move the object in different positions. An experimental analysis would be needed for acoustic trapping. Such as the stability, and the distribution of the acoustic radiation forces, which is to model the acoustic radiation forces against pressure. Some mathematical approaches are also required for the levitation force and sound pressure calculations.

Theory

The fundamental principle of acoustic levitation is that the angular frequency of the acoustic wave would be generated by the bottom emitter with high power and made by the transducer(the area of the transducer would be designed usually with a large area of radiation, in order for the stability of the levitation).[21] Then the wave would be emitted by the bottom emitter to the reflector surface, the acoustic standing wave would be constructed between the transducer and reflector at a certain distance in the vertical direction.

Acoustic radiation force

In single-axis acoustic levitation, a standing wave is generated by two transducers facing each other. The standing wave consists of alternating high-pressure and low-pressure zones that create an acoustic radiation force, which can levitate small objects. The acoustic radiation force is a result of the interaction between the standing wave and the object being levitated. As the sound waves pass through the object, they create small vibrations that cause the object to move. The direction and magnitude of the movement are determined by the phase and amplitude of the sound waves at each point on the object's surface. At the points where the standing wave is at maximum pressure, the object experiences a force pushing it toward the center of the levitation zone. At the points where the standing wave is at minimum pressure, the object experiences a force pushing it away from the center. By carefully adjusting the frequency and amplitude of the sound waves, it is possible to create a stable levitation zone where the acoustic radiation force is balanced and the object is suspended in mid-air. The strength of the acoustic radiation force depends on several factors, including the frequency and amplitude of the sound waves the size and shape of the object, and the distance between the object and the transducers. In general, larger objects require stronger sound waves to levitate, while smaller objects can be levitated with weaker waves. The pioneer of using acoustic radiation force on the levitated particle was first contributed by King[24], who investigated the acoustic radiation pressure on the spherical particle in the frictionless medium by applying the stationary pressure field. He attempted to determine the average resultant pressure on the rigid spherical particle, which comes up with highly accurate results for acoustic pressure. Afterward, there are a large number of research studies focusing on the radiation force levitating spherical particles in the medium of rigid, fluid, compressible, and plane waves, in order for the progress of acoustic levitators and other types of levitations[25].

To begin with the acoustic radiation force, the basic formulation for fluid dynamics would be considered as the continuity equation[27]:

$$\frac{\partial \rho}{\partial t} + \Delta \cdot (\rho u) = 0 \qquad\qquad eq[2].$$

The second term of the equation is the multiplication of the mass density and the velocity of the fluid, which is called the Navier-Stokes equation for momentum conservation. Regardless of the viscous force, the Navier-Stokes equation becomes

$$\rho\left[\frac{\partial u}{\partial t} + (u \cdot \Delta)u\right] = -\Delta p \qquad\qquad eq[3].$$

For the sake of the complexity of nonlinear relations, the perturbation method would be needed to solve the equation. And then the second-order term for the fluid field would be taken into account,

$$p = p_{atm} + p_1 + p_2$$
$$u = u_1 + u_2$$

$$\rho = \rho_0 + \rho_1 + \rho_2$$

Where $P_1 = c_0^2 \rho_1$, the c_0 is the speed of the sound in the fluid[33].

Assuming it is time-harmonic fields,

$$\rho_1 = \rho_1(r)e^{-i\omega t},$$
$$p_1 = p_1(r)e^{-i\omega t},$$
$$v_1 = v_1(r)e^{-i\omega t},$$

Velocity potential ϕ_1,

$$v_1(r) = \Delta\phi_1(r),$$
$$p_1(r) = i\rho_0\omega\phi_1(r),$$
$$\rho_1(r) = i\frac{\rho_0\omega}{c_0^2}\phi_1(r).$$

The potential works for the wave equation:

$$\Delta^2\phi_1 = \frac{1}{c_0^2}\partial_t^2\phi_1 = -\frac{\omega^2}{c_0^2}\phi_1. \qquad\qquad \text{eq[4].}$$

Then Navier-Stoke equation in the second order would be obtained

$$\Delta p_2 = -\rho_0\frac{\partial u_2}{\partial t} - \rho_1\frac{\partial u_1}{\partial t} - \rho_0(u_0 \cdot \Delta)u_1 \qquad\qquad \text{eq[5].}$$

Then the procedure reached the equation for acoustic radiation pressure

$$< p_2 > = \frac{1}{2}\kappa_0 < p_1^2 > - \frac{\rho_0}{2} < u_1^2 > \qquad\qquad \text{eq[6].}$$

Where κ_0 is compressibility of the fluid [33], which is given by

$$\kappa_0 = \frac{1}{(\rho_0 c_0^2)}$$

Then it came up with the time average <fg> for two harmonic fields

$$< fg > = \frac{1}{2}Re[f(r)g^*(r)]$$

When the radius of the rigid object is smaller than the wavelength, the density and the compressibility of the fluid would be treated as weak point scatterers for acoustic waves, then first-order scattering theory could be introduced for the first order acoustic field ϕ_1.

The first-order acoustic field is given by the incident field and scattered field:

$$\phi_1 = \phi_{in} + \phi_{sc}$$
$$v_1 = \Delta\phi_1 = \Delta\phi_{in} + \Delta\phi_{sc}$$
$$p_1 = i\rho_0\omega\phi_1 = i\rho_0\omega\phi_{in} + i\rho_0\omega\phi_{sc}.$$

Now the acoustic radiation force on the rigid object can be written as

$$F_{rad} = -\int_{S_0} dS\{< p_2 > \hat{n} + \rho_0 < (n \cdot v_1)v_1 >\}$$

$$= -\int_{dS} dS\{[\frac{1}{2}\kappa_0 < p_1^2 > - \frac{1}{2}\rho_0 < v_1^2 >]\hat{n} + \rho_0 < (n \cdot v_1)v_1 >\} \qquad \text{eq[7.1].}$$

This relation shows that the force carries out over the whole surface of the sphere by integration.

Then according to Gor'kov's potential relation, for the sphere with a radius smaller than the wavelength in the acoustic field, the acoustic radiation potential is given by

$$U = 2\pi R^3 [\frac{f_1}{3\rho_0 v_0^2} < (p_1^{in})^2 > - \frac{f_2\rho_0}{2} < (u_1^{in})^2 >] \qquad \text{eq[7.2]}.$$

Where p_1^{in} and u_1^{in} indicate the incident field of the particle, and $f_1 = 1 - \frac{\rho_0 v_0^2}{\rho_m v_m^2}$, and $f_2 = 2(\frac{\rho_m - \rho_0}{2\rho_m + \rho_0})$, the ρ_m and v_m are the sphere density and the speed of the sound.

Gor'kov potential would help to solve the acoustic pressure and incident velocity field. First, the acoustic pressure in the standing wave can be written as

$$p_1^{in} = p_0 cos(kz)cos(\omega t)$$

And the incident velocity field is given as

$$u_1^{in} = \frac{p_0}{\rho_0 v_0} sin(kz)sin(\omega t)$$

According to the trigonometry, the $< cos^2(\omega t) >=< sin^2(\omega t) >= 1/2$. Then the potential is obtained as

$$U_{rad} = \frac{p_0^2 \pi R^3}{\rho_0 v_0^2} [\frac{1}{3}cos^2(kz) - \frac{1}{2}sin^2(kz)] \qquad \text{eq[8]}.$$

Finally, the acoustic radiation force can be obtained

$$F_{rad} = \frac{5\pi R^3 k p_0^2}{6\rho_0 v_0^2} sin(2kz). \qquad \text{eq[9]}.$$

Acoustic radiation pressure dominates in the acoustic levitation, which was discovered by Rayleigh on the subject of nonlinear radiation force study. This relation is derived from the adiabatic equation of state[28]

$$p = P - P_0 = A(\frac{\rho - \rho_0}{\rho_0}) + \frac{B}{2}(\frac{\rho - \rho_0}{\rho_0})^2$$

Then obtained the relation of Euler equation for non-viscous Newtonian fluid

$$\rho \frac{\partial v}{\partial t} + \rho(v \cdot \Delta)v = - \Delta p \qquad \text{eq[10]}.$$

Where P_0 is the ambient pressure, $P = P_0 + p$, in which the sum of the acoustic pressure p and ambient density ρ_0. ρ is the sum of the acoustic density ρ' and ambient density. The nonlinearity parameters A and B have equations which are $A = \rho_0 c_0^2, B = (\Gamma - 1)\rho_0 c_0^2$, in the cases of an ideal gas, the term, where γ stands for a specific heat ratio. Since this relation has been investigated by several researchers to produce different results, turns out the modified equation for Rayleigh radiation pressure in the ideal gas became

$$P_{ra} = \frac{1+\gamma}{2}(1 + \frac{sin(2kh)}{2kh}) < E > \qquad \text{eq[11]}.$$

Where E is the energy density of the wave, which can be written as

$$< E >= (a_0^2/4)(\rho_0\omega^2/sin^2 kh). \qquad \text{eq[12]}.$$

Therefore the energy can be obtained by the angular frequency ω of the wave, vibration amplitude, a_0 and the distance h is from vibration source to the target point. K stands for the wavenumber $k=\omega/c$. Since this method has one limitation on the studies of plane waves, it would require Eulerian coordinates in three dimensional, rather than the Lagrangian coordinates in 1 dimension. The reason for this is that the levitated object would experience sound on the surface, and the mean pressure is related to Lagrangian in 1 dimension, therefore, in this case, the mean pressure in 3-D used Eulerian coordinates. The relation for the mean pressure in Eulerian coordinates is

$$<P^E - P_0> = <V> - <K> + C \qquad \text{eq[13]}.$$

Where V and K are the potentials and the kinetic energy density of the wave. They are given by

$$V = \frac{1}{2}\frac{1}{\rho_0 c^2}p^2$$

$$K = \frac{1}{2}\rho_0 v^2$$

Thus the relation of mean pressure we obtained is the acoustic radiation pressure in the standing wave levitation.

The radiation force for an object with a rigid sphere could be obtained as

$$F = -\frac{5}{6}\pi\rho_0|A|^2(kR_s)^3 sin(2kh) \qquad \text{eq[14]}.$$

This relation for acoustic radiation force is obtained by King on the investigation of spherical objects levitated in the fluid, which can be found in the source[29].

It is necessary to calculate the acoustic field produced by the acoustic emitter, and the forces generated in the field. The acoustic pressure would stand as a complex scalar field if a single frequency is generated in the frequency domain. When the models execute in the time range, the pressure would be a real scalar field. The total field would be explained theoretically as the pressure distribution of the waves affected by all the parameters. The incident field is also required to be considered when the acoustic field is generated somewhere with no reflectors to affect the emitted wave in the domain. Therefore the total field would be the sum of the incident field and scattered field.

The incident acoustic field related to the object can be expressed as spherical harmonics expansion[22]

$$p_i(r) = \sum_{n=0}^{\infty} \sum_{m=-n}^{n} j_n(kr)S_n^m Y_n^m(\theta,\varphi), \qquad \text{eq[15]}.$$

Where $j_n(kr)$ stands for the spherical Bessel functions with n-th order, and the k is the wavenumber and $k = \omega/c$, c is the speed of the sound, ω is the angular frequency. The vector r of the spherical particle can be expressed mathematically in Cartesian space

$$x = rsin(\theta)cos(\varphi)$$
$$y = rsin(\theta)sin(\varphi)$$
$$z = rcos(\theta)$$

Where r is the Euclidean distance and can be written as $r = \sqrt{x^2 + y^2 + z^2}$, the polar angle θ equals to arccos(z/r) with the interval $[0,\pi]$. The azimuth angle $\varphi = atan2(y,x)$ and $\varphi \in (-\pi,\pi]$.

The equation for spherical harmonic bases Y_n^m can be expressed as

$$Y_n^m(\theta, \varphi) = \sqrt{\frac{2n+1}{4\pi}\frac{(n-m)!}{(n+m)!}} P_n^m(cos\theta)e^{im\varphi}, \qquad \text{eq[15.1].}$$

Where P_n^m is the associated Legendre polynomials and it is corresponding to the Legendre polynomials $P_n(x)$, which is

$$P_n^m(x) = (-1)^m(1-x^2)^{m/2}\frac{d^m}{dx^m}P_n(x) \qquad \text{eq[15.2].}$$

Similarly, the scattered acoustic field can be expressed as

$$P_s(r) = \sum_{n=0}^{\infty} \sum_{m=-n}^{n} h_n(kr)\hat{S}_n^m Y_n^m(\theta, \varphi) \qquad \text{eq[15.3].}$$

Where $h_n(kr) = j_n(kr) + iy_n(kr)$ are the spherical Hankel functions with summing up n-th terms, and $y_n(kr)$ are the spherical Bessel functions with n-th order. As the calculations of scattered and incident waves are accomplished in the frequency domain, the model is not dependent on the time delays for the spherical particle moving with speed in the limited domain. Thus there are two cases when the spherical particle is compressible or in expansion. When the spherical particle is compressible, the equation for coefficients $\hat{S}_n^m = c_n S_n^m$ is expressed as

$$\hat{S}_n^m = -\frac{j_n(k_0a)j'_n(k_pa)-\underline{Z}j'_n(k_0a)j_n(k_na)}{h_n(k_0a)j'_n(k_pa)-\underline{Z}h'_n(k_0a)j_n(k_pa)}S_n^m \qquad \text{eq[15.4].}$$

Where a stands for the radius of the sphere, and $\underline{Z} = (\rho_p c_p)/(\rho_0 c_0)$ refers to the relative impedance. On the contrary, the expended spherical particle for every single transducer has been shown as

$$S_n^m = \sum_j S_n^m = \sum_j \frac{1}{j_n(kr_0)}\int_\Gamma p^j(r)Y_n^m(\theta, \varphi)^* d\Gamma \qquad \text{eq[15.5].}$$

In order to generate an acoustic field, the approach is called free-field piston source, which is the simplest method for far-field levitating. It takes advantage of fast manipulation. First of all, before calculating the acoustic field, the complex acoustic pressure generated by a single frequency needs to be considered and can be calculated as

$$P(r) = P_0 V \frac{D_f(\theta)}{d}e^{i(\varphi+kd)} \qquad \text{eq[16].}$$

Where the P_0 stands for the efficiency output of a single transducer. V stands for the excitation signal amplitude of crests. d is the distance in free space. K is the wavenumber, φ emitting phase, and λ wavelength. D_f can be simplified and written as sinc(kasinθ). Then the total acoustic field P for N transducers would be $P = \sum_{j=1}^{N} P_j$.

After the particle is trapped in the acoustic field, the trapped particle would experience acoustic radiation force. Gork'ov potential can be used to describe the force acting on the particle. This method is usually applied for the particle smaller than the wavelength, which can be described as

$$F = -\Delta U \qquad \text{eq[17].}$$

Where the potential U can be written as incident pressure

$$U = 2K_1(|p|^2) - 2K_2(|p_x|^2 + |p_y|^2 + |p_z|^2) \qquad \text{eq[18].}$$

$$K_1 = \frac{1}{4}V(\frac{1}{c_0^2\rho_0} - \frac{1}{c_s^2\rho_s})$$

$$K_2 = \frac{3}{4}V\left(\frac{\rho_0 - \rho_s}{\omega^2 \rho_0 (\rho_0 + 2\rho_s)}\right)$$

Based on the Gor'kov's method for calculating the acoustic radiation force by introducing potential, the method can be used to determine the particle smaller than the wavelength of sound.

Radiation Flux integral

Radiation force exerted on the particle could be determined by the integral over momentum fluxes with respect to the spherical area. Then the second-order approximation of the radiation force is given as

$$F = \iint_S \left\{ \left(\tfrac{1}{4}\rho|v|^2 - \frac{1}{4\rho c^2}|p|^2\right)n - \tfrac{1}{2}\rho Re[v^*(v \cdot n)]\right\}dS \qquad \text{eq}[19].$$

Where p is the complex pressure and v is the velocity of the particle caused by the acoustic field with respect to the time dependence exp(-iwt)., S is the surface, n is perpendicular to the surface of the system. This reference can be viewed from [30]. Therefore, the acoustic force can be expressed as the equation

$$F = \iint_S -\frac{1}{2\rho c^2}<P^2> +\tfrac{1}{2}\rho<v^2> -\rho<(vn)\cdot v>da \qquad \text{eq}[20].$$

Where $<P^2>$ is the expectation value of pressure in the time domain.

Review

Studies on the acoustic radiation force have been done by many groups of researchers to study the radiation force exerted on either regular or irregular particles. [25] this paper showed the work on investigating acoustic radiation force on the solid sphere in the Rayleigh regime by considering larger mass density than that in the fluid medium. They used mathematical equations for Gork'ov potential function and the boundary condition for the acoustic analysis. The results are also obtained from the method of using Gauss and Bessel standing waves to a harmonic oscillation, which came up with the conclusion that the inaccuracy of the natural oscillation frequency would be caused by the gravitational force. The experiment used six solid spheres made up of different materials with radiuses of 1.5mm, then they used a high-speed camera to record the oscillation of the vibration. They observed that the levitated spheres would be in simple harmonic motion surrounding the equilibrium position after the wire stopped holding those spheres, which is affected by the radiation forces. In this study, the work is useful for investigating the acoustic radiation force on the spherical particles.The calculation of acoustic radiation forces in three dimensions would be slightly different; it requires spherical coordinates on the levitated spherical particle. The study[26] showed the calculation of the elastic sphere in three dimension radiation force by applying the scattering beam to a sphere. Furthermore, they showed the measurements for the radiation stress tensor with superposition on the incident field and scattered field. They came to an agreement with Gor'kov on the fact that the radiation force exerted on smaller spherical particles would be greater on a standing plane wave. The results of the work they obtained for the radiation force exerted on the small spheres using helicoidal Bessem beams is that the 25-degree angle with helicity 1 requires the force to be zero at the equilibrium position. The maximum force was obtained as $-3.3 10^{-12} N$ for the transverse displacement of 0.35 times the wavelength. The result is lying to the agreement to King's work, that the sphere would approach the pressure nodes while being easily collected in multiple equilibrium positions. Furthermore, the work also showed the axial and transverse stability in the Bessel beam by applying the Legendre polynomials and by defining three half-cone angles with degrees 50, 60, and 70. The forces in the x and y directions are coherent with the force in the z-direction.

There are two types of levitation systems, one is standing wave levitation and squeeze film levitation.[28]In this article, the researchers demonstrated two types of levitation techniques, which are standing wave levitation and squeeze film levitation. The tiny objects would be held stably in the position near the pressure node or the antinode of the standing wave field formed by the radiator and reflector. The levitated objects would be required to have a size smaller than the sound wave in order to accomplish the levitation. [28] On the other hand, the squeeze film levitation levitates objects with more than half a wavelength higher than the emitter. In order to accomplish stable levitation, the standing wave would be required in the middle of the radiator and the targeting object. They introduced the idea of a radiator and reflector to levitate materials in a noncontact way for standing wave levitation. Standing wave acoustic levitation is the simplest method to levitate particles in mid-air. For example, It can be applied to space to lift objects with different materials by using sound waves. An acoustic chamber can also be applied to the standing wave levitation to levitating materials. The squeeze film levitation is used to levitate large planer objects, the levitated object would be close to the radiation surface with an altitude smaller than the wavelength of the standing wave applied to the levitation. In this paper[28], they used a mathematical approach to model the sound field by introducing a sound beam in a cylindrical coordinate system with the Bessel function. The applied method in their experiment is the squeeze-film levitation with the reflector and radiator, and the laser interferometer is used to detect the position of the reflector. The object used for the levitating test is a compact disc with a mass of 16g. The result shows that the stable levitation would be obtained at a position slightly higher than 0.5λ. The simulation result for the radiation forces at different positions shows that as the distances of the levitation increase, the radiation forces would decrease to zero. In addition, the simulation result for the radiation forces versus the distance is slightly higher than the actual measurement. The visualization of the radiation pressure against the frequency shows that the radiation pressure would decrease when the frequency is less than 20kHz and following that the radiation pressure would rise up slightly. Therefore, concluded the fact that the radiation pressure would not be affected by the frequency.

[27]In this review, Jackson (2021) demonstrated that stable equilibrium points could be created by some analysis approaches for acoustic levitation. Then the results proved a contradiction to the theoretical analysis. Furthermore, they studied fluid dynamics and came to an agreement with the experimental results. The article started with the 1 dimension analysis of acoustic standing waves with respect to the pressure.

The pressure can be expressed in 1 dimension as

$$p(z,t) = p_0 cos(kz)cos(\omega t)$$
eq[21].

Where the k is the wavenumber, ω is the angular frequency proportional to the period T. p_0 is the maximum amplitude of the standing wave.

Forces in the x and y axis would be balanced due to the symmetry, therefore, the z-direction force would be the only case that needed to be considered. There would be only one force caused to the pressure field resulting from the non-gravity condition. Therefore the equation of motion governed by the cubic-shaped particle can be expressed by[27]

$$mz = 4a^2[p(z-a,t) - p(z+a,t)],$$
$$= 4a^2p_0\{cos[k(z-a)] - cos[k(z+a)]\}cos(\omega t),$$
$$= 8a^2p_0 sin(ka)sin(kz)cos(\omega t)$$

The resultant force on the particle would be reduced to zero when it satisfies the condition ka=nπ. Therefore one could conclude that the force caused by a standing wave depends on the size of the particle. Jackson(2021) [27] stated that they would use the size of the particle as 2a < λ for the investigation of the experiment. There are three factors to be considered. One is the time average, in which the average velocity would vanish when the cosine function approaches zero. The conclusion of it is that the particle would keep the same location at the initial location. The second point obtained from the equation is that the zero force can be analyzed from the equilibrium position when the sine function is equal to zero.

By presenting the equation of motion acting on the spherical particle with radius R.

$$8a^2p_0 sin(ka) -> \frac{4\pi p_0}{k^2}[sin(kR) - kRcos(kR)].$$

Zheng (2018) [32] explained that the experiment analysis for dimensionless acoustic levitation force and restoring force are corresponding to the geometric parameters, they investigated the air particle suspension position by determining the potential under half wavelength. They came up with the result that the largest sound pressure would occur in the middle. The levitated ball would be depressed by the effect of the acoustic pressure. They concluded that the levitation force on the rigid ball would be affected by the sound pressure levels. The lowest level of sound pressure would line up with the suspension of the rigid ball. The results conclude that the stability of the sound suspension would be affected by the increased sound pressure.

In the study of investigating the stability of spherical and ellipsoidal using acoustic levitation, Foresti [34] with his co-workers introduced Computational fluid dynamics (CFD) and the finite-volume method (FVM) in the acoustic levitation study. CFD method is for the measurement of radiation pressure. FVM is a method for determining the acoustic radiation pressure on the cylinder by setting a standing wave as a boundary condition. Additionally, the finite-element method (FEM) model and finite-volume method (FVM) for observing the forces on the spherical and ellipsoidal particles in a different size range in an axisymmetric acoustic levitation medium. They considered three parameters in the analysis, which are particle axial position, two different ranges of particle size that have larger particles (radius/wavelength of sound > 0.25), the standing wave would not be valid, and radial force on the particle would be less than 0. Smaller particles (radius/ wavelength of sound <0.01) result in a viscous force reducing the stability of the levitation. The last factor is the particle shape; results showed that the balance of the axial and radial forces is affected by the radial force to the stability, this is in terms of the ellipsoidal particles. In the situation of inviscid, the FEM model would be needed, as it can be applied to a 3-dimensional model to calculate the inviscid medium problem. The advantages of inviscid case computation are first, that it is much faster than the viscous study with more accurate results on determining small particles. The CFD model can be applied to study viscosity. In the study of viscosity using the CFD model, the dynamic mesh (DM) can be introduced with frequency for sinusoidal axial motion and velocity on the emitter. For the sake of the accuracy of the radiation force on the greater objects, DM needs a first-order discrete time step. The data showed two results for the radius/wavelength of sound=0.1730 and =0.1153 under the driving frequency 40kHz, the redshift appears on the axial position.

The FEM model is used by considering different lengths horizontally in the levitation. The data showed that the radial force decreases to negative when the height is at the peak(H/λ=0.73). In this case, the levitation under such radial force would not be accomplished due to the instability. In the case of stable area, the (H/λ=0.9, radius/λ<0.25), The radial force would stand positive. The root means squared velocity produces the maximum radiation pressure force in this range. At the unstable area(H/λ=0.73, R/λ=0.25), the radial force is at the negative region, and the levitated particle would be in a compressed state. Foresti et al.(2012)[34] states that the largest particle size being levitated is approximately 0.25λ, which is corresponding to the particle shape. Foresti et al.(2012)[34] also indicated some effects on the viscosity of the CFD model. The effect of medium viscosity is considered the fact that it would reduce the stability of the levitated particle. In terms of the smaller particle, the R/λ greater than 0.0125, the results of viscosity increasing would cause a decrease in axial force, RMS pressure, and velocity. The convincing observation shows that increased viscosity reduces radial force more than that axial force, which could affect the velocity of the ellipsoid horizontally. The balanced relation between radial and axial forces could affect the stability of the ellipsoid particles.

The equations for obtaining the FEM model in terms of space and time could be derived as

$$-\Omega^2 \frac{1}{K_f}\underline{P_a} - \Delta \cdot (\frac{1}{\rho}\Delta \underline{P_a}) = 0$$

The first term and the second term can be obtained by the oscillate harmonic with an angular frequency and inviscid fluid at the equilibrium condition.

The CFM model considers the relation for a conserved momentum to derive with stress tensor. The equation can be derived by the energy transferred to conduction and viscous dissipation, which can be written as

$$\frac{\partial}{\partial t}(\rho E) + \Delta \cdot (\underline{v}(\rho E + p)) = \Delta \cdot (k \cdot \Delta T + \underline{\tau} \cdot \underline{v})$$

The equation for energy is

$$E = \frac{v^2}{2} - \frac{P}{\rho} \qquad \text{eq[22].}$$

In order to predict the sound field in the acoustic levitation, the method is called finite difference time domain (FDTD), which is remarkable for determining the sound field. This method can be found in the work[35]. The acoustic-FDTD method for the longitudinal sound waves is performed in the article with discussed results. They used a cylinder for the longitudinal acoustic wave with an x-y spatial model. They applied a stress tensor and Hooke's law for the two-dimensional differential equations. The results show that the waveforms detected in the cylindrical samples would perform normalization conditions caused by the maximum amplitude.

Methods

Acoustic levitation experiment requires to measure the acoustic radiation force. There are several approaches to measure the acoustic radiation force in single-aixs levitation. Laser Doppler vibrometry, interferometry, and optic modulation. Interferometry is the simplest method to measure the acoustic radiation force. First of all, the single-axis acoustic levitation system needs to be constructed to make sure that the object is levitated. Place a mirror or reflective surface near the levitated object, so that it reflects the acoustic wave from the transducer. Then use an interferometer, such as a Michelson interferometer, to measure the displacement of the reflective surface due to the acoustic radiation force. Next, record the interference pattern produced by the interferometer, which reflects the phase shift and amplitude modulation of the reflected light due to the displacement of the reflective surface. Then analyze the interference pattern to extract the displacement, velocity, and acceleration of the reflective surface, and use this information to calculate the acoustic radiation force acting on the levitated object. Repeat the measurement for different frequencies, amplitudes, and positions of the ultrasonic transducer, in order to characterize the acoustic radiation force and optimize the levitation conditions.

To measure the radial and axial forces in a single-axis acoustic levitation system, the combination of sensors and signal processing techniques can be used. There are several ways of measuring the radial and axial forces in a single-axis acoustic levitation system, specialised sensors would be needed that are capable of measuring forces in multiple directions. First one is piszoelectric force sensors, which are capable of measuring static and dynamic forces in three directins. They work by converting mechanical stree into an electrical charge, which can be measured using specialised electronic equipment. To measure axil and radial forces, the force sensor needs to be positioned so that it is aligned with the direction of the force that requires to be measured. Laser Doppler vibrometry is a non-contact technique for measuring the displacement and velocity of an object. It works by shining a laser beam on the object and measuring the Doppler shift in the reflected light, which is proportional to the object's velocity. By integrating the velocity over time, the object's displacement can be obtained and the forces acting on it. Laser Doppler vibrometry(LDV) is particularly useful for measuring radial forces, which can be difficult to measure using other tchiques. Acoustic emissions can be used to indirectly measure the forces acting on an object in an acoustic levitation system. Acoustic emissions are high-frequency sound waves that are generated by the interaction of the levitated object with the standing wave. Signal processing techniques can be used, which are Fourier analysis or wavelet analysis, to extract the frequency components of the signals obtained from the LDV and the axial displacement sensor. This would allow the radial and axial components of the displacement and velocity signals to be separated. The calibration procedure correlates the signals from the force sensor, the LDV, and the axial displacement sensor with the actual forces and displacements acting on the levitated object. This calibration should be done under controlled conditions, such as varying the input voltage to the ultrasonic transducer, and it should be repeated periodically to ensure accuracy and reliability. The feedback control system adjusts the frequency and amplitude of the ultrasonic transducer to maintain stable levitation of the object. This control system should use the signals from the force sensor, the LDV, and the axial displacement sensor to compute the required adjustments, and it should be designed to handle disturbances and changes in the system parameters.

Piezoelectric load cells are commonly used as high precision force sensors to measure the total force acting on a levitated object in acoustic levitation experiments. These load cells rely on the piezoelectric effect, which generates an electrical charge in response to mechanical stress or strain. To measure the total force acting on the levitated objected, there are several steps for using peizoelectric load. First thing is to choose a suitable piezoelectric load cell with the appropriate range, sensitivity, and accuracy. mount the load cell in a suitable location that allows it to measure the force actng on the levitated object. Connect the load cell to a computer, that can record and analyze the output signal from the load cell. Perform a calibration procedure to correlate the output signal from the load cellwith the actual forces acting on the levitated object. This may involve applying known weights or forces to the load cell and recording the corresponding output signal, and then fitting a calibration curve to the data. Then use calibrated load cell to monitor the total force acting on the levitated object during the acoustic levitation. This can help to ensure stable and accurate levitation, and to detect any changes or perturbations in the levitation system.

Analysis and Discussion

Acoustic levitation is a process in which an object is suspended in mid-air by utilizing sound waves instead of using physical contact or conventional methods of suspension. It has been used for various applications such as conducting experiments with non-following samples, for the isolation of fragile objects in industrial processes, and for even the propulsion of certain materials. The data analysis of acoustic levitation has been conducted in order to understand its underlying principles and to improve its efficiency and effectiveness. An important factor in the analysis of acoustic levitation is the speed of sound in the medium employed. This plays an important role in the synthesis of sound waves and hence in the levitation process.

The speed of sound in air is approximately 343 m/s, while in water it is 1482 m/s. Acoustic levitation is also affected by various parameters such as the frequency, power, intensity, and direction of the sound waves.

There are several analysis results for acoustic levitation. Acoustic levitation is based on the formation of standing waves, which are produced by two opposing sound waves that interfere with each other. The nodes and antinodes of the standing wave create regions of high and low pressure that can trap objects. The frequency and wavelength of the sound waves used in acoustic levitation are important parameters that affect the size and stability of the levitated objects. Higher frequencies and shorter wavelengths are generally more effective for leaving smaller objects. Particle size and shape of the objects being levitated also play a critical role in interfering with the stability of the levitation. Spherical objects are generally easier to levitate than irregularly shaped objects, and larger objects require more energy to levitate than smaller ones. Material properties being levitated can also affect the levitation stability. Such as the viscosity and surface tension of liquids can affect their ability to form stable levitation patterns. Experimental design is also a factor that can affect the results of the experiment, the placement of the levitated object can impact the stability and behavior of the levitation.

There are several basic parameters for the acoustic system, which are the frequency for each of the transducers is 40kHz, speed of the sound is 343 meters per second. The sound pressure amplitude is 100000.0Pa. And the density of the air is 1.2kg/m^3, and the radius of the spherical object is 0.5×10^{-3}m. Then, the wavelength and the wavevector can be calculated by the relations.

$$\lambda = \frac{c}{f},$$ eq[23].

where c is the speed of sound, and f is the frequency.

$$k = 2 \times \frac{\pi}{\lambda}.$$

Following by the resonant frequency of the spherical object can be determined.

$$f = \frac{c}{2\times\pi\times a}.$$

By using simple programming code on python, calculations for the resonant frequency of the spherical objects can be obtained with value 1.1×10^5Hz, the wavenumber k is $732.7\frac{1}{m}$. wavelength is 0.0086m. The volume of the sphere is $5.24\times 10^{-10}m^3$. The levitation force can be obtained from the equation in terms of the amplitude and radius of the spherical object, then the levitation force is 11.2N.

The sound pressure field and levitation force can be manipulated using python, and the plot of the standing wave field for the position (m) against sound pressure amplitude (Pa)

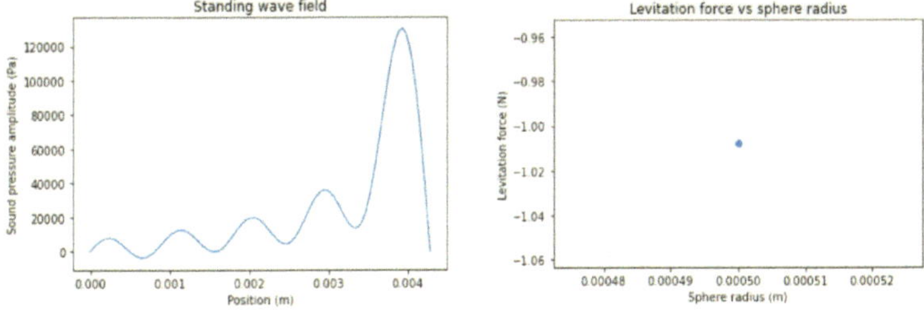

Figure1: the first graph on the left-hand side shows the plot of the standing wave field for the relation of Position to the sound pressure amplitude. The graph on the right-hand side shows the relation between the levitation force with the spherical radius.

The graph of a standing wave field shows the amplitude of the sound waves at different points in space. The graph shows the amplitude of the sound waves along the axis of the levitation zone in the single-axis acoustic levitation system. At points where the sound waves are at maximum pressure, the amplitude of the wave would be also high. At points where the sound waves are at minimum pressure, the amplitude of the wave would be low. This pattern at regular intervals along the standing wave. The amplitude of the sound waves is increased, and the amplitude of the standing wave field would also increases. This means that the high-pressure zones would be higher, and the low-pressure zones would be lower. In other words, the difference between the maximum and minimum pressures would be greater. The position of the high-pressure and low-pressure zones in the standing wave field is determined by the frequency of the sound waves. As the frequency increases, the distance between the high-pressure and low-pressure zones decreases, and the distance between the high-pressure and low-pressure zones decreases. This means that the size of the levitation zone decreases, but the acoustic radiation force becomes stronger.

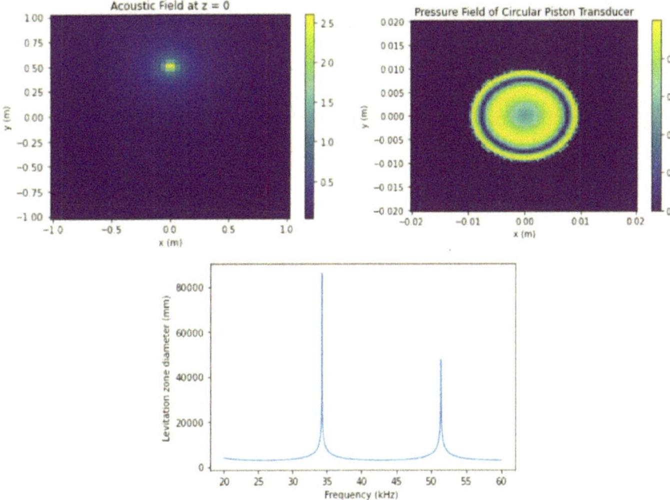

Figure 2: the left-hand side of the graph shows the mesh grid plot of the acoustic field at z=0 on the axis of the x-y plane, and the graph on the right-hand side shows the mesh grid plot of pressure field of the circular piston Transuder on the x-y plane. The third graph shows the relation between frequency(KHz) and levitation zone diameter(mm)

The graph shows on the left-hand side the acoustic field at z=0 in the z-y plane, which is essential in an acoustic levitation system since the levitation zone is typically located. In a single-axis acoustic levitation system, the levitation zone is created by two transducers facing each other. The transducers emit sound waves that interfere with each other to create a standing wave, as discussed earlier, this standing wave consists of alternating high-pressure and low-pressure zones, which can be visualized as a complex pattern of peaks and troughs in the acoustic field. At z=0, the acoustic field can be analyzed in the x-y plane to determine the position and shape of the levitation zone. The acoustic field in this plane will consist of a series of high-pressure and low-pressure zones, which would be distributed in a regular pattern that is determined by the frequency and phase of the sound waves. The position and shape of the levitation zone can be controlled by adjusting the frequency and phase of the sound waves emitted by the transducers. The pattern of high-pressure and low-pressure zones in the x-y plane at z=0 is determined by the geometry of the transducers and the wavelength of the sound waves. In general, the size of the levitation zone would decrease as the frequency of the sound waves increases, and the spacing between the high-pressure and low-pressure zones would also decrease.

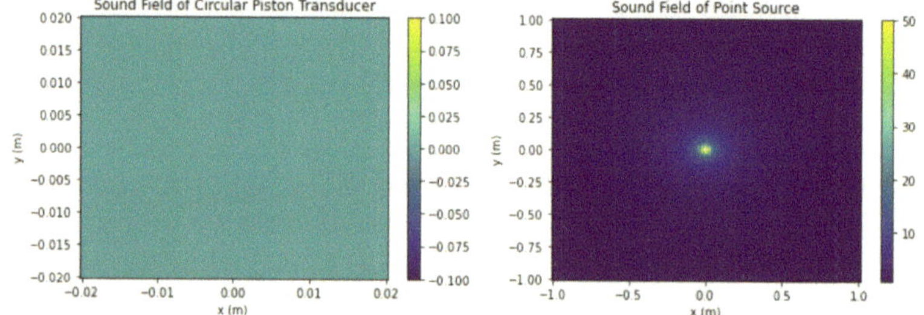

Figure 3: the mesh grid plot for the sound field of the circular position Transducer is shown on the left side and the mesh plot for the sound field Point source is shown on the right-hand side.

Figure 3 shows the sound field of a circular piston transducer. The sound field of a circular piston transducer can be described by the pressure field it generates. The pressure field is the distribution of the sound pressure levels in the medium around the transducer. The sound pressure levels are a measure of the intensity of the sound field at different locations. The pressure field generated by a circular piston transducer can be calculated using theoretical models or numerical simulations. The pressure field is affected by a number of factors, including the frequency of the sound wave, the size and shape of the transducer, and the properties of the medium. In general, the pressure field generated by a circular piston transducer would have a maximum value at the center of the transducer and would decrease in intensity as the distance from the transducer increases. The shape of the pressure field would depend on the geometry of the transducer and the frequency of the sound wave. Additionally, the sound field of the circular piston transducer can also be described in terms of the particle velocity field and the particle displacement field. The particle velocity field decreases the motion of particles in the medium in response to the sound wave, while the particle displacement field describes the displacement of particles from their equilibrium positions.

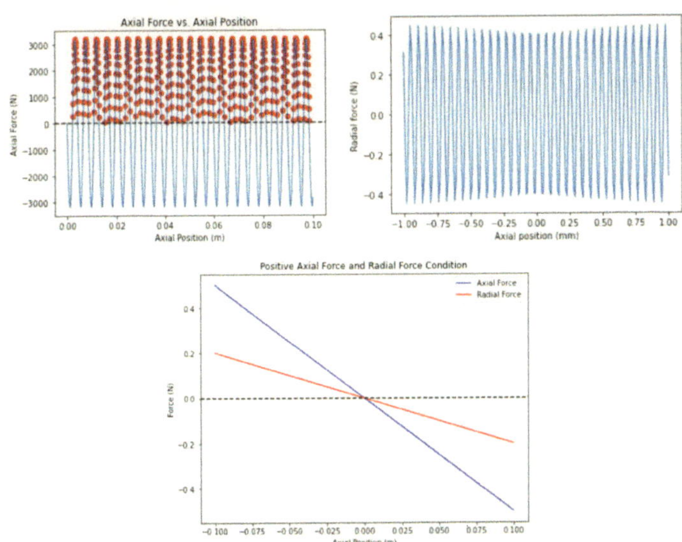

Figure 4: those two graphs are the axial force against the axial position on the left-hand side, this plot shows axial force vs. axial position, the marked red dots are showing the positive axial force, the horizontal line at y=0 indicates the boundary condition between positive and negative axial force. and the radial force against the axial position on the right-hand side, note that in the region above the y-axis when y=0, the radial force is positive, which means that the levitation is at a stable position. The third plot shows the axial position versus axial force and radial force, the red line is radial force, while the blue line is axial force. In this case, both the radial force and axial force are positive, which indicates stable levitation.

In single-axis levitation, the axial force and axial position are important parameters for providing information about the stability of the levitation system. The axial force is the force acting along the axis of the acoustic beam and is proportional to the gradient of the sound pressure field. In acoustic levitation, the axial force is responsible for levitating the particle and keeping it centred along the acoustic beam. If the axial force is in the negative region, it indicates that the particle is experiencing a force in the opposite direction of the desired levitation. In other words, the acoustic radiation force is pushing the particle downwards instead of upwards. This can be due to various reasons such as improper positioning of the transducer or the particle being too heavy or large for the given acoustic frequency. In order to reach stable levitation, the axial force should be positive and within a certain range. By measuring the axial force, it is possible to monitor the stability of the levitation system and make adjustments as needed to maintain stable levitation. The axial position is the position of the particle along the axis of the acoustic beam, and is related to the phase of the sound wave, in acoustic levitation, the axial position is an important parameter that can affect the stability of the levitation system. If the particle is not properly centered along the acoustic beam, it may experience an unbalanced acoustic radiation force that can cause it to drift out of the levitation zone. The radial force can be measured using a force sensor or a balance beam. A negative radial force and axial position indicate that the particle is being pulled toward the center of the acoustic field. This is because the negative radial force is acting inwards, towards the center of the field, and the negative axial position indicates that the particle is located below the equilibrium position. When a particle is located below the equilibrium position, the axial force acts upwards, opposing the gravitational force. However, the radial force acts inwards, pulling the particle toward the center of the field. As a result, the particle may oscillate around the equilibrium position, but it would not be stable levitation. To stably levitate the particle, both the radial and axial forces must be positive, meaning that the particle is pushed away from the center of the field and towards the equilibrium position. Figure 4 shows the axial force against the axial position on the left-hand side and the radial force versus the axial position on the right-hand side. The plot shows that the axial force above the y-axis is indicated as red dots, which means that the in this region the axial force is positive and is in stable levitation. While on the right-hand side, the plot shows the radial force and axial position on the right-hand side, the region above the horizontal axis y=0, the radial force is positive and the levitation is stable. When both the axial force and radial force are positive, then the particle could be levitated stably at the equilibrium position. In the plot of axial position versus axial force and radial force in the position range from -1 to 1, both the radial and axial forces are positive, which indicates stable levitation.

There are some limitations in single-axis levitation, one of the major limitations is that it can only levitate small objects, generally in the size range of millimetres. This is due to the fact that the size of the levitation zone is directly proportional to the frequency of the sound wave used, and higher frequencies result in smaller levitation zones. In addition, the stability of the levitated object decreases as its size increases. Another limitation is that the levitation is affected by external factors such as air currents, temperature gradients, and vibrations, which can cause the object to move out of the levitation zone and become less stable. Finally, single-axis levitation is not suitable for irregularly shaped objects levitation, since the acoustic radiation force is affected by the object's size, shape, and density.

Conclusion

Both optical trapping of aerosol and single-axis acoustic levitation is contactless methods for manipulating small particles and objects. Optical trapping of aerosol uses beams to create a force that traps particles in a stable region, while single-axis acoustic levitation uses sound waves to create a standing wave field that traps objects at the nodes of the wave. Both techniques have their advantages and limitations. Optical trapping of aerosol has the advantage of being able to trap and manipulate particles in a non-destructive manner, with high precision and control. However, it is limited to particles that are within a certain size range and may be affected by factors such as particle shape, refractive index, and laser power. Single-axis acoustic levitation has the advantage of being able to levitate and manipulate small objects that are not transparent and is less affected by environmental factors such as air currents and temperature gradients. However, it is limited in its ability to levitate larger or irregularly shaped objects.

References

[1]*Nanostructured potential of optical trapping using a plasmonic ...* (no date). Available at: https://www.researchgate.net/publication/236098812_Nanostructured_Potential_of_Optical_Trapping_Using_a_Plasmonic_Nanoblock_Pair (Accessed: December 16, 2022).

[2]*Optical trapping and binding - iopscience - institute of physics* (2013). Available at: https://iopscience.iop.org/article/10.1088/0034-4885/76/2/026401 (Accessed: December 16, 2022).

[3]*High-performance reconstruction of microscopic force fields from Brownian trajectories*, *Nature News*. Nature Publishing Group. Available at:https://www.nature.com/articles/s41467-018-07437-x (Accessed: December 6, 2022).

[4]M. D. Summers, D. R. Burnham, and D. McGloin, "Trapping solid aerosols with optical tweezers: A comparison between gas and liquid phase optical traps," Opt. Express 16, 7739-7747 (2008)

[5] Neuman, K.C. and Block, S.M. (2004) *Optical trapping, The Review of scientific instruments*. U.S. National Library of Medicine. Available at: https://www.ncbi.nlm.nih.gov/pmc/articles/PMC1523313/ (Accessed: December 5, 2022Pérez García, L. *et al.* (2018)

[6]*High-performance reconstruction of microscopic force fields from Brownian trajectories, Nature News*. Nature Publishing Group. Available at:https://www.nature.com/articles/s41467-018-07437-x (Accessed: December 6, 2022).

[7]*Investigation of dust grains by optical tweezers for space applications* (2022). Available at:https://www.researchgate.net/publication/364162908_Investigation_of_dust_grains_by_optical_tweezers_for_space_appli cations (Accessed: December 6, 2022).

[8]H;, S.A.B.A.D.J.R.-D. (2021) *Enhanced Signal-to-noise and fast calibration of optical tweezers using single trapping events, Micromachines*. U.S. National Library of Medicine. Available at: https://pubmed.ncbi.nlm.nih.gov/34067843/ (Accessed: December 15, 2022).

[9]Ashkin, A. *et al.* (1986) *Observation of a single-beam gradient force optical trap for dielectric particles, Optics Letters*. Optica Publishing Group. Available at: https://opg.optica.org/abstract.cfm?URI=ol-11-5-288 (Accessed: December 3, 2022).

[10]*Optical control and characterisation of aerosol - researchgate* (2009). Available at: https://www.researchgate.net/publication/244137645_Optical_control_and_characterisation_of_aerosol (Accessed: December 26, 2022).

[11]Di Leonardo, R., Ruocco, G., Leach, J., Padgett, M.J., Wright, A.J., Girkin, J.M., Burnham, D.R. and McGloin, D. (2007). Parametric Resonance of Optically Trapped Aerosols. *Physical Review Letters*, [online] 99(1). doi:10.1103/physrevlett.99.010601.

[12]Bouloumis, T.D. and Nic Chormaic, S. (2020) *From far-field to near-field micro- and nanoparticle optical trapping, MDPI*. Multidisciplinary Digital Publishing Institute. Available at: https://www.mdpi.com/2076-3417/10/4/1375#B12-applsci-10-01375 (Accessed: December 20, 2022).

[13]*T-matrix method for modelling optical tweezers - taylor & francis* (2010). Available at: https://www.tandfonline.com/doi/full/10.1080/09500340.2010.528565 (Accessed: December 14, 2022).

[14]P. W. Barber & H. Massoudi (1982) Recent Advances in Light Scattering Calculations for Nonspherical Particles, Aerosol Science and Technology, 1:3, 303-315, DOI: 10.1080/02786828208958596

[15]*Optical trapping of dielectric particles in arbitrary fields* (2000). Available at: https://www.researchgate.net/publication/12018248_Optical_trapping_of_dielectric_particles_in_arbitrary_fields (Accessed: December 27, 2022).

[16]Ling, L. *et al.* (2010) *Optical forces on arbitrary shaped particles in optical tweezers, AIP Publishing*. American Institute of PhysicsAIP. Available at: https://aip.scitation.org/doi/pdf/10.1063/1.3484045?casa_token=ZOC63ugdteEAAAAA%3AzsyXK07aU03Vsnh9iAiAka0BGgj eGie4ThrNdkM7e9yXR1eMItKP1Rc6i9oLP1aoX0aEkNIKXBs (Accessed: December 27, 2022).

[17]Loke, V.L.Y., Nieminen, T.A., Heckenberg, N.R. and Rubinsztein-Dunlop, H. (2009). -matrix calculation via discrete dipole approximation, point matching and exploiting symmetry. *Journal of Quantitative Spectroscopy and Radiative Transfer*, [online] 110(14-16), pp.1460–1471. doi:10.1016/j.jqsrt.2009.01.013.

[18]Fushimi, T. , Hill, T. L., Marzo, A., & Drinkwater, B. W. (2018). Nonlinear trapping stiffness of mid-air single-axis acoustic levitators. Applied Physics Letters, 113(3), [034102]. https://doi.org/10.1063/1.5034116

[19]Marzo, A., Barnes, A. and Drinkwater, B.W. (2017). TinyLev: A multi-emitter single-axis acoustic levitator. *Review of Scientific Instruments*, 88(8), p.085105. doi:10.1063/1.4989995.

[20]Andrade, MAB, Bernassau, A, Okina, FTA & Adamowski, JC 2017, 'Acoustic levitation of large objects in air', Paper presented at 24th International Congress on Sound and Vibration, London, United Kingdom, 23/07/17 - 27/07/17.

[21]Zang, Y., Chang, Q., Wang, X., Su, C., Wu, P. and Lin, W. (2022). Natural oscillation frequencies of a Rayleigh sphere levitated in standing acoustic waves. *The Journal of the Acoustical Society of America*, 152(5), pp.2916–2928. doi:10.1121/10.0015142.

[22]Zehnter, S., Andrade, M.A.B. and Ament, C. (2021). Acoustic levitation of a Mie sphere using a 2D transducer array. *Journal of Applied Physics*, 129(13), p.134901. doi:10.1063/5.0037344.

[23]Pazos Ospina, J.F., Contreras, V., Estrada-Morales, J., Baresch, D., Ealo, J.L. and Volke-Sepúlveda, K. (2022). Particle-Size Effect in Airborne Standing-Wave Acoustic Levitation: Trapping Particles at Pressure Antinodes. *Physical Review Applied*, 18(3). doi:10.1103/physrevapplied.18.034026.

[24]King, L.V. (1934). On the Acoustic Radiation Pressure on Spheres. *Proceedings of the Royal Society of London. Series A, Mathematical and Physical Sciences*, [online] 147(861), pp.212–240. Available at: https://www.jstor.org/stable/pdf/96275.pdf [Accessed 6 Feb. 2023].

[25]Zang, Y., Chang, Q., Wang, X., Su, C., Wu, P. and Lin, W. (2022). Natural oscillation frequencies of a Rayleigh sphere levitated in standing acoustic waves. *The Journal of the Acoustical Society of America*, 152(5), pp.2916–2928. doi:https://doi.org/10.1121/10.0015142.

[26]Baresch, D., Thomas, J.-L. and Marchiano, R. (2013). Three-dimensional acoustic radiation force on an arbitrarily located elastic sphere. *The Journal of the Acoustical Society of America*, 133(1), pp.25–36. doi:https://doi.org/10.1121/1.4770256.

[27]Jackson, D.P. and Chang, M.-H. (2021). Acoustic levitation and the acoustic radiation force. *American Journal of Physics*, 89(4), pp.383–392. doi:https://doi.org/10.1119/10.0002764.

[28]Zhao, S. and Wallaschek, J. (2009). A standing wave acoustic levitation system for large planar objects. *Archive of Applied Mechanics*, 81(2), pp.123–139. doi:https://doi.org/10.1007/s00419-009-0401-3.

[29]King (1934). On the acoustic radiation pressure on spheres. *Proceedings of the Royal Society of London. Series A - Mathematical and Physical Sciences*, 147(861), pp.212–240. doi:https://doi.org/10.1098/rspa.1934.0215.

[30]Marzo, A. (2020). Standing Waves for Acoustic Levitation. *Acoustic Levitation*, pp.11–26. doi:https://doi.org/10.1007/978-981-32-9065-5_2.

[31]Research, U. (n.d.). *Acoustic Levitator*. [online] Instructables. Available at: https://www.instructables.com/Acoustic-Levitator/.

[32]Zhang, F. and Jin, Z. (2018). The Experiment of Acoustic Levitation and the Analysis by Simulation. *OALib*, [online] 05(10), pp.1–8. doi:https://doi.org/10.4236/oalib.1104948.

[33]Bruus, H. (2012). Acoustofluidics 7: The acoustic radiation force on small particles. *Lab on a Chip*, 12(6), p.1014. doi:https://doi.org/10.1039/c2lc21068a.

[34]Foresti, D., Nabavi, M. and Poulikakos, D. (2012). On the acoustic levitation stability behaviour of spherical and ellipsoidal particles. *Journal of Fluid Mechanics*, [online] 709, pp.581–592. doi:https://doi.org/10.1017/jfm.2012.350.

[35]Tanikaga, Y., Takizawa, T., Sakaguchi, T. and Watanabe, Y. (n.d.). *A Study on analysis of intracranial acoustic wave propagation by the finite difference time domain method 43.35 Wa Biological effects of ultrasound, ultrasonic tomography*. [online] Available at: http://www.conforg.fr/acoustics2008/cdrom/data/fa2002-sevilla/forumacusticum/archivos/ultgen005.pdf [Accessed 21 Feb. 2023].

YOUR KNOWLEDGE HAS VALUE

- We will publish your bachelor's and
  master's thesis, essays and papers

- Your own eBook and book -
  sold worldwide in all relevant shops

- Earn money with each sale

Upload your text at www.GRIN.com
and publish for free